MW00874944

Kenley's Line Plot Graph

Another Math Adventure

Kathleen L. Stone

Enjoy these other books by Kathleen L. Stone

Penguin Place Value
A Math Adventure

Number Line Fun
Solving Number Mysteries

Riley the Robot
An Input/Output Machine

Mason the Magician
Hundreds Chart Addition

Katelyn's Fair Share Picnic
More Math Fun

Money Tree Mysteries
Adventures with Quarters

From My Quilted Heart to Yours
*Heart Warming Quilts and Heart Healthy Recipes for
Your Loved Ones*

Copyright © 2015 Kathleen L. Stone

All rights reserved.

ISBN–13: 978-1508572978
ISBN-10: 1508572976

Dedicated with love to the newest member of our family, our beautiful granddaughter, Kenley! I love you with all my heart!

There's someone I'd like you to meet
Down at the Elma Zoo.
She's a great mathematician
Named Kenley the Kangaroo.

Kenley and her friends,
A monkey and two giraffes,
Spend their days counting
And making line plot graphs.

This afternoon Kenley's counting
All the feet passing by her spot.
Let's see if we can help her
Create a new line plot.

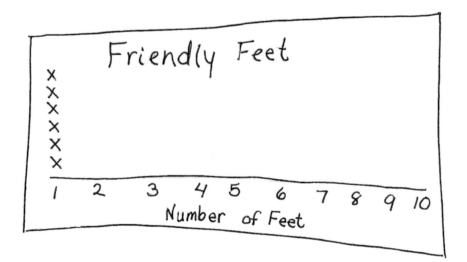

Friendly Feet

x
x x
x x x
x x x x
x x x x x

1 2 3 4 5 6 7 8 9 10
Number of Feet

Six snails are in the garden
So what should Kenley put
In the column that shows us
Animals with one foot?

Walking by her pen
Are *three* zoo keeper staff.
Where will Kenley mark
This data on her graph?

Kenley marked *two*
For animals with *four* feet
When she saw Eric and Elsie
Walking down the street.

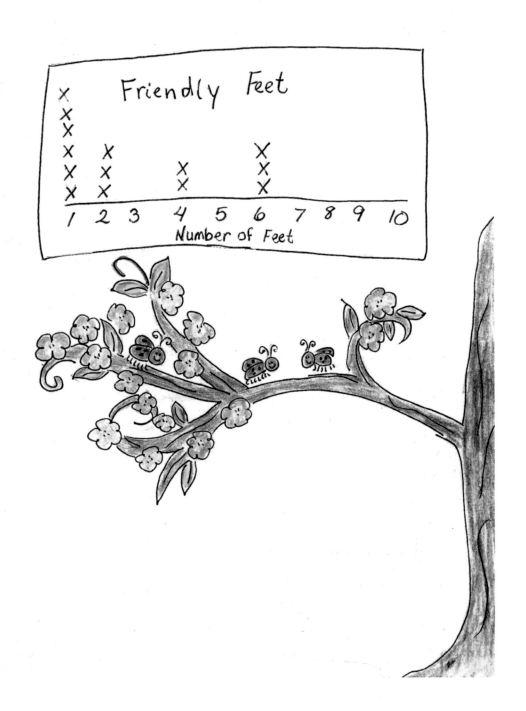

Can you figure out why
For *six* feet Kenley marked a *three*?
Here's a little hint,
Look carefully at the tree.

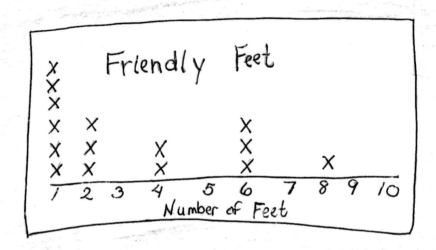

Friendly Feet

```
X
X
X
X        X                        X
X        X        X               X
X        X        X               X                 X
—————————————————————————————————————————————————————————
1    2    3    4    5    6    7    8    9    10
              Number of Feet
```

An octopus in the aquarium
Is searching for a snack to eat.
What should Kenley mark
For animals with *eight* feet?

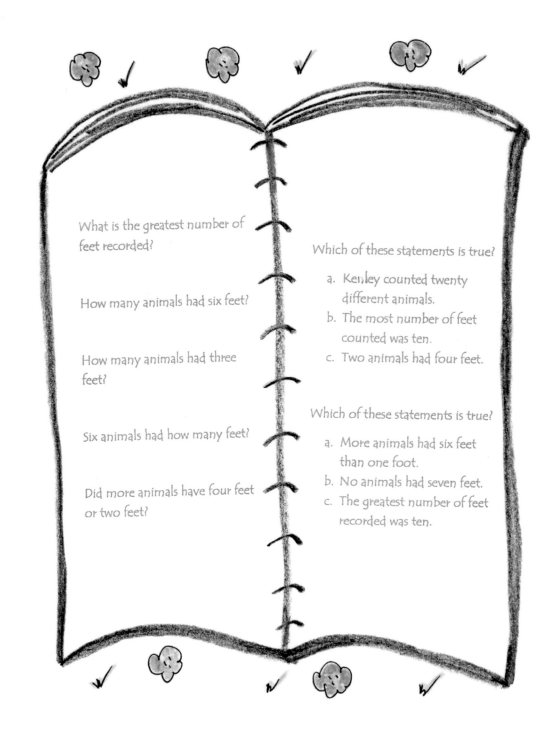

What is the greatest number of feet recorded?

How many animals had six feet?

How many animals had three feet?

Six animals had how many feet?

Did more animals have four feet or two feet?

Which of these statements is true?

a. Kenley counted twenty different animals.
b. The most number of feet counted was ten.
c. Two animals had four feet.

Which of these statements is true?

a. More animals had six feet than one foot.
b. No animals had seven feet.
c. The greatest number of feet recorded was ten.

Kenley's graph is finally finished
But take another look.
Are you able to answer
The questions in her book?

Line Plot Graphs

A line plot is a special type of graph that shows frequency of data along a horizontal scale or number line. Line plots provide a quick and easy way for children to organize and interpret data. Line plot graphs are a great way to introduce probability and provide practice with additional math skills, such as measurement.

Children should be given many opportunities to work with concrete examples (hands on manipulatives) before moving on to more abstract concepts when working with line plots.

Using this Book

Post-It notes provide a handy way to cover the data in the illustrations as you read *Kenley's Line Plot Graph* aloud. This way you can discuss what Kenley's line plot should look like and then check answers with the illustrations. I have my students create their own line plots as we read and see if their graphs match Kenley's.

Enrichment Activities

What's My Lucky Number?
Materials needed:

regular six-sided die
How to Play

Children roll their die fourteen times and record the number rolled each time on their line plot graph:

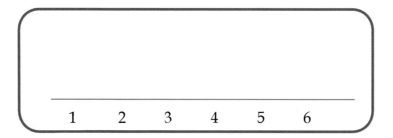

Once their line plot is complete, ask the following questions:
- ♥ Which number did you roll the most?
- ♥ How many times did you roll the number 4?
- ♥ What was the number you rolled the least number of times?

Mystery Bag

Materials needed:

bag(s) that contain a variety of objects to measure
ruler(s)

How to Play

Children may work individually or in small groups measuring, to the nearest inch or centimeter, a variety of objects that are in their bag. They will use this measurement data to create a line plot. You can extend this activity by having them create their own questions about their line plots.

Additional Line Plot Topics …
- ♥ number of goals each player scored
- ♥ length of feet (inches or centimeters) in classroom or family
- ♥ number of pets each student has
- ♥ height (to closest inch or cm) of group of friends
- ♥ number of teeth lost
- ♥ spelling test scores
- ♥ bedtime of you and your friends

ABOUT THE AUTHOR

Kathleen Stone is a National Board Certified educator and is currently teaching second grade. She loves spending time with her family. She and her husband Gary live in the Olympia area. When not teaching, Kathleen can often be found quilting, sitting by the lake reading, or exploring nearby parks with her grandchildren!

Math is all around us
No matter where you turn
Open your mind to the wonders of math
And all that you can learn

Made in the USA
Monee, IL
16 June 2021

71443469R00017